# Embracing the Future of hybrid and technology

## Exploring the World of Electronic and Hybrid Vehicles

By

# Justin  cools

# Table of content

# Introduction

In a rapidly evolving world, the concept of embracing the future holds immense significance. As technological advancements continue to shape our lives and societies, we find ourselves standing at the intersection of uncertainty and incredible potential. Embracing the future entails a proactive mindset, fueled by adaptability and forward-thinking ideas, to navigate the challenges that lie ahead while capitalizing on the opportunities that await us. Whether it be in the realms of science, business, education, or even personal growth, a willingness to embrace the future paves the way for innovation, progress, and ultimately, a brighter tomorrow. By exploring the multitude of possibilities that lie before us, we can harness the power of change, spark transformation, and forge new paths that will redefine the world as we know it. The time has come for us to embrace the future with open arms and embark on a journey of growth and exploration that will shape our collective destiny.

# Chapter 1

# The Rise of Electronic and Hybrid Vehicles

Revolutionizing the Automotive Industry In recent years, the automotive industry has been experiencing a significant shift towards sustainable transportation. The rise of electronic and hybrid vehicles has paved the way for a cleaner, greener, and more environmentally friendly mode of transportation. This chapter delves into the emergence of electronic and hybrid vehicles, highlighting their impact on the automotive industry, their benefits, and the challenges they face.

Evolution of Electronic and Hybrid Vehicles: The concept of electric vehicles dates back to the early 19th century, but it wasn't until the late 20th century that they gained traction. - Major milestones in the commercialization of electric vehicles, such as the launch of Tesla's Roadster in 2008, kicked off the industry. Advancements in battery technology, cost reduction, and increased charging infrastructure have Understanding Hybrid Vehicles Hybrid vehicles have gained significant popularity in recent years due to their energy efficiency and environmental benefits. Let's dive deeper into understanding these vehicles and why they are considered a great alternative to traditional gasoline-powered cars.

definition of hybrid vehicles: A hybrid vehicle is a type of automobile that utilizes both an internal combustion engine (typically gasoline-powered) and an electric motor. These vehicles combine the advantages of both gasoline engines and electric motors, resulting in increased fuel efficiency and reduced emissions.

How Hybrid Vehicles Work: Hybrid vehicles operate on a unique principle called regenerative braking. When the vehicle decelerates or brakes, instead of wasting this energy as heat, it is used to charge the electric motor's battery. As a result, the electric motor assists the combustion engine, reducing the overall load on the gas engine, which means less fuel consumption.

## A Sustainable Transportation Solution:

A Sustainable Transportation Solution: Advancing Towards a Greener Future In today's world, finding sustainable transportation solutions is crucial to combating climate change, reducing air pollution, and promoting a greener future. This article explores various sustainable transportation options and initiatives that are transforming the way we travel, making transportation cleaner, greener, and more efficient.

Electric Vehicles (EVs): Electric vehicles have gained significant traction as a sustainable transportation choice. They generate no tailpipe emissions, thereby reducing air pollution and reducing the need for fossil fuels.EVs are becoming more affordable and have longer ranges, and the charging infrastructure is expanding rapidly. Governments and car manufacturers are increasingly investing in EVs, aiming for mass adoption and a sustainable transportation future.

Public Transportation:  Efficient and accessible public transportation plays a pivotal role in sustainable mobility. High-speed trains, light rail systems, buses running on biofuels, and other low-emission alternatives can help reduce the number of private vehicles on the road.

**Understanding the need for sustainable transportation options**

Sustainable transportation options are vital for several reasons, including:

Environmental Benefits:  They reduce air pollution and greenhouse gas emissions, helping combat climate change and improve air quality.

Energy Efficiency:  Sustainable transportation modes like public transit and electric vehicles are more energy-efficient, reducing resource consumption.

Reduced Congestion: Promoting walking, cycling, and public transit can alleviate traffic congestion, save time, and reduce stress.

Health Benefits: Active transportation options promote physical activity, reducing the risk of health issues like obesity and heart disease. Economic Advantages: Sustainable transportation investments can create jobs and stimulate local economies.

# Equity and Accessibility:

They provide affordable and accessible transportation options for all, reducing transportation-related disparities.

Resilience: Sustainable transportation is more resilient to disruptions like fuel shortages or extreme weather events. Overall, sustainable transportation is crucial for a healthier environment, improved public health, and enhanced quality of life.

## Exploring the environmental benefits of electronic and hybrid vehicles

The study by the Keck School of Medicine and its lead author, health sciences at the Keck School of Medicine, has found that local changes in electric vehicle (EV) adoption can improve the health of communities. The researchers found that while total ZEVs increased over time, adoption was slower in low-resource zip codes, referred to as the "adoption gap." This disparity presents an opportunity to restore environmental justice in communities disproportionately affected by pollution and related health problems.

Benefits for health and the climate: The research team analyzed and compared four different datasets to study the effects of electric vehicle adoption. They obtained data on ZEVs from the California Department of Motor Vehicles, U.S. Environmental Protection Agency air monitoring sites, and the percentage of adults in each zip code who held bachelor's degrees. Every 20 more zero-emission vehicles (ZEVs) per 1,000 people has resulted in a modest decrease in $NO_2$ levels and a 3.2% decrease in emergency visits linked to asthma. This pattern is more noticeable in zip codes with lower levels of education, indicating that underprivileged regions might gain from ZEVs. Due to sustainability and environmental concerns, the automobile industry is moving toward more environmentally friendly vehicles. This transition is being driven by the desire for electric vehicles and the expanding market for such vehicles.

Environmental Concerns:  The foremost reason behind the surge in demand for eco-friendly cars is the increasing awareness of environmental issues. Concerns about climate change, air pollution, and carbon emissions have prompted consumers to seek cleaner transportation alternatives. Eco-friendly cars, including electric vehicles (EVs) and hybrid vehicles, offer reduced emissions and a smaller carbon footprint, aligning with the global push for sustainability.

Government Incentives: Many governments around the world have implemented incentives and regulations to encourage the adoption of eco-friendly vehicles. These incentives can include tax credits, rebates, and access to carpool lanes, making eco-friendly cars more appealing to consumers. Additionally, stricter emissions standards and penalties for high-emission vehicles have further accelerated the shift toward cleaner options.

Technological Advancements: Advances in battery technology have significantly improved the performance and affordability of electric vehicles. Longer driving ranges, faster charging times, and lower costs have made EVs a viable choice for a broader range of consumers. As technology continues to evolve, eco-friendly cars are becoming more accessible and practical.

 Rising Fuel Costs:Fluctuating fuel prices have a direct impact on consumer choices. With the volatility of oil prices, the stability of electricity rates for EVs is appealing to many. Eco-friendly cars not only reduce greenhouse gas emissions but also offer potential long-term cost savings on fuel and maintenance.

**Consumer Preferences:** A shift in consumer preferences is evident, with many individuals now valuing sustainability and eco-consciousness when making purchasing decisions. Automakers are responding by offering a wider range of eco-friendly models to cater to this growing demand. As eco-friendly cars become more stylish and high-performing, they appeal to a broader demographic.

 Infrastructure Development:The expansion of charging infrastructure for electric vehicles is a critical factor in the growing market demand for eco-friendly cars. Governments and private companies are investing heavily in building charging networks, alleviating the "range anxiety" often associated with EVs and making them more convenient for consumers.

Corporate Responsibility: Numerous companies are actively pursuing their corporate social responsibility and sustainability objectives.As a result, corporate fleets are increasingly turning to eco-friendly cars to reduce their environmental impact. This trend not only demonstrates commitment to sustainability but also sets an example for employees and customers alike.

Global Market Trends:The global market for eco-friendly cars is interconnected. As major automakers invest in eco-friendly technology and expand their offerings, consumers worldwide are exposed to a broader range of choices. This global trend amplifies the market demand for eco-friendly cars as consumers seek options that align with international sustainability standards. In conclusion, the growing market demand for eco-friendly cars is a multifaceted phenomenon driven by environmental concerns, government incentives, technological advancements, economic factors, and shifting consumer preferences. As the automotive industry continues to evolve, eco-friendly cars are poised to play an increasingly significant role in shaping the future of transportation.

# Chapter 2

## Powering the Future:

Hybrid vehicles represent a significant step toward a more sustainable and efficient transportation future. These innovative vehicles combine traditional internal combustion engines with electric propulsion systems, offering a range of benefits that contribute to a cleaner and more energy-efficient future.

Reduced Environmental Impact:Hybrid vehicles typically produce fewer emissions compared to their conventional counterparts. By combining the power of electric motors with gasoline or diesel engines, hybrids can operate more efficiently and release fewer harmful pollutants into the atmosphere. This reduction in emissions is crucial for combating air pollution and addressing climate change.

 Improved Fuel Efficiency:Hybrids are designed to optimize fuel efficiency. Their electric motors assist the internal combustion engine during acceleration and provide power during idle or low-speed driving, reducing the overall consumption of fossil fuels. This not only saves drivers money at the pump but also decreases our reliance on finite fossil fuel resources.

 Regenerative Braking: Hybrid vehicles often feature regenerative braking systems that capture and store energy that is usually wasted as heat during braking. This recovered energy is then used to recharge the vehicle's battery, further enhancing its efficiency and extending the distance it can travel on electric power alone.

Incentives and Rebates: Many
governments around the world offer
incentives and rebates to encourage the
adoption of hybrid vehicles. These incentives
can include tax credits, reduced registration
fees, and access to carpool lanes, making
hybrid ownership even more appealing to
consumers.

Advancements in Technology: As
technology continues to evolve, hybrid
vehicles are becoming more sophisticated
and capable. Newer models often have longer
all-electric ranges and improved overall
performance, making them a practical choice
for a broader range of drivers.

Charging Infrastructure: The growth of
hybrid vehicles is driving investments in
charging infrastructure. While not purely
electric, hybrids still rely on battery
technology, and as the number of hybrid
vehicles on the road increases, so does the

demand for charging stations, which benefits both hybrid and electric vehicle owners.

 Transition to All-Electric: Hybrid technology is a stepping stone towards the broader adoption of all-electric vehicles. By easing consumers into the world of electric propulsion, hybrids help familiarize people with the benefits of electric driving and reduce range anxiety. This gradual transition paves the way for a future where electric vehicles dominate the automotive landscape. In conclusion, the future of hybrid vehicles is promising. They play a vital role in reducing emissions, conserving fuel, and transitioning to a more sustainable transportation ecosystem. As technology continues to advance, hybrid vehicles will become even more efficient and accessible, contributing to a greener and more eco-friendly future for transportation.

## Technology and Innovations in Electronic and Hybrid Vehicles Revolutionizing the Road:

Technology and Innovations in Electronic and Hybrid Vehicles" In recent years, the automotive industry has witnessed a remarkable transformation with the advent of electronic and hybrid vehicles. These innovative technologies have not only changed the way we drive but have also ushered in a new era of sustainability and efficiency on our roads. Let's delve into some of the key advancements that are driving this automotive revolution:

Battery Technology:  At the heart of electronic and hybrid vehicles lies their battery technology. Manufacturers are constantly pushing the boundaries to develop batteries that offer greater energy density and a longer lifespan. Lithium-ion batteries have been the go-to choice, but research is

ongoing to explore alternatives like
solid-state batteries, which promise even
higher energy density and safety.

Electric Motors:  Electric motors are
becoming more efficient and compact.
Permanent magnet motors and induction
motors are commonly used, but innovations
in motor design and materials are enhancing
their performance. These motors provide
instant torque, resulting in quicker
acceleration and a smoother driving
experience.

Regenerative Braking: Hybrid vehicles
employ regenerative braking systems that
capture and store energy during braking. This
stored energy is then used to power the
vehicle, improving overall efficiency.
Advances in regenerative braking technology
are making hybrids even more fuel-efficient.

Power Electronics: Power electronics play a crucial role in converting electricity from the battery to a form usable by the vehicle's motor. Innovations in power electronics have led to more efficient conversion processes, reducing energy losses, and extending the range of electric vehicles (EVs).

Charging Infrastructure: To support the growth of EVs, charging infrastructure has seen significant advancements. High-speed DC fast chargers are becoming more widespread, reducing charging times dramatically. Moreover, wireless charging technology is being explored, allowing for convenient and cable-free recharging.

Range Extenders: Some hybrid vehicles incorporate range extenders like small gasoline engines or fuel cells. These serve as backup power sources when the battery is depleted, extending the vehicle's range significantly.

 Connected and autonomous features: Electronic and hybrid vehicles are increasingly equipped with advanced driver-assistance systems (ADAS) and connectivity features. These include adaptive cruise control, lane-keeping assist, and autonomous driving capabilities, making transportation safer and more convenient.

 Materials and Lightweighting: Manufacturers are utilizing lightweight materials such as carbon fiber and aluminum to reduce the overall weight of vehicles. This enhances efficiency and extends battery life, contributing to a greener footprint.

 Sustainable Materials: Eco-friendly materials are gaining prominence in vehicle manufacturing. Recycled plastics, sustainable interior fabrics, and eco-conscious production processes are being embraced to reduce the environmental impact of vehicle production.

Energy Management Systems: Advanced software and algorithms are employed to optimize energy consumption in electronic and hybrid vehicles. These systems intelligently manage power distribution between the battery, motor, and other components to maximize efficiency. As technology continues to advance, electronic and hybrid vehicles are poised to play a pivotal role in reducing greenhouse gas emissions and combating climate change. These innovations not only benefit the environment but also promise a future of quieter, cleaner, and more efficient transportation for all. In conclusion, the landscape of electronic and hybrid vehicles is constantly evolving, driven by relentless innovation. From battery technology to connected features, these advancements are shaping a greener and smarter future for the automotive industry. As consumers, we can anticipate even more exciting developments in the years to come, ultimately transforming the way we travel.

## Examining the advancements in battery technology and energy storage systems

In recent years, there have been significant advancements in battery technology and energy storage systems, driven by the increasing demand for renewable energy sources, electric vehicles, and portable electronic devices. These developments are transforming various industries and have the potential to reshape our energy landscape. Let's delve into some key aspects of these advancements:

Lithium-ion Dominance: Lithium-ion batteries are a leading energy storage technology due to their high energy density, longer cycle life, and low self-discharge rates. Researchers are continuously improving their performance through new materials and manufacturing processes.

Solid-state batteries: Replacing liquid electrolytes with solid electrolytes, are gaining attention from companies like Toyota and Samsung for potential increased energy density, safety, and lifespan.

Energy Density Improvements: Researchers are working on increasing the energy density of batteries, making them smaller and more efficient. This is crucial for electric vehicles (EVs) to extend their range and reduce charging times.

Fast Charging:  Faster charging solutions are being developed, enabling EVs and other devices to recharge in minutes rather than hours. Companies like Tesla and Porsche have introduced fast-charging networks.

Grid-Level Storage:  Energy storage systems are increasingly used for grid-level applications to store excess renewable energy for later use. This helps balance the intermittent nature of renewable sources like solar and wind.

Second-Life Batteries: Used EV batteries, which may no longer meet the demands of a car, can find a second life in energy storage systems. This sustainable approach reduces waste and costs.

Materials Innovation: Advances in materials science are crucial for battery improvements. Silicon anodes, for example, could significantly boost capacity, while research into new cathode materials aims to reduce costs and increase performance.

AI and Battery Management: Artificial intelligence is being utilized to enhance battery performance. AI algorithms can predict battery degradation, improve charging efficiency, and extend battery lifespan.

Environmental Concerns: As battery production scales up, there is growing scrutiny of the environmental impact. Recycling and responsible disposal of batteries are becoming increasingly important.

Market Growth: The global energy storage market is booming, with numerous startups and established companies competing to innovate and capture a share of this rapidly expanding industry. These advancements in battery technology and energy storage systems hold great promise for a more sustainable and electrified future. They are essential not only for the widespread adoption of renewable energy but also for the continued development of electric vehicles and the overall reduction of carbon emissions.

## Discussing the integration of electric motors and hybrid powertrains:

Integrating electric motors into hybrid powertrains is a pivotal development in the automotive industry. This integration combines the benefits of traditional internal combustion engines with electric propulsion, resulting in more efficient and environmentally friendly vehicles. One of the primary advantages of hybrid powertrains is their ability to optimize energy usage. Electric motors can assist the internal combustion engine during acceleration, reducing fuel consumption and emissions. They can also act as generators during deceleration, recovering energy that would otherwise be wasted as heat and storing it in the vehicle's battery for later use. Additionally, hybrid powertrains allow for various driving modes, such as full electric, hybrid, and regenerative braking. This flexibility offers drivers the choice to prioritize fuel efficiency, performance, or zero-emission driving, depending on their needs and preferences. The integration of electric motors and hybrid powertrains also plays a crucial role in reducing greenhouse gas emissions and meeting stricter emissions standards. Many automakers are investing heavily in hybrid technology as a transitional step toward fully electric vehicles, as it allows them to significantly decrease their carbon footprint while maintaining the convenience of traditional refueling infrastructure. Moreover, electric motors provide instant torque, enhancing the overall performance and responsiveness of hybrid vehicles. This combination of improved efficiency and better driving dynamics makes hybrid cars an attractive option for a wide range of consumers. In conclusion, the integration of electric motors into hybrid powertrains represents a significant step forward in the automotive industry's pursuit of sustainability and efficiency. As technology continues to advance, we can expect further innovations in this field, leading to even more capable and environmentally friendly hybrid vehicles.

## Exploring the use of regenerative braking and other energy-saving features:

Regenerative braking and energy-saving features are important topics in various industries, including automotive and transportation. Here's some content exploring their use:

Regenerative Braking in Electric Vehicles: Regenerative braking is a crucial feature in electric vehicles (EVs). It allows the vehicle to recover and store energy during braking, which can then be used to extend the vehicle's range. This technology not only increases energy efficiency but also reduces wear and tear on traditional braking systems.

Energy-Saving Features in Buildings: Beyond transportation, energy-saving features are essential in building design and construction. Smart lighting, heating, and cooling systems, as well as energy-efficient insulation and windows, play a significant role in reducing energy consumption in homes and commercial buildings.

Regenerative Braking in Trains and Public Transit: Regenerative braking is also employed in trains and public transit systems. Trains can capture energy during braking and redistribute it to other cars or the grid. This reduces energy costs and the environmental impact.

Renewable Energy Integration: The energy saved through regenerative braking and other energy-saving features can often be integrated into renewable energy systems, such as solar or wind power. This synergy promotes sustainable energy usage and reduces reliance on fossil fuels.

Environmental Benefits:  Exploring these technologies is crucial for reducing greenhouse gas emissions and combating climate change. By minimizing energy waste and optimizing energy use, we can mitigate the environmental impact of various industries.

Challenges and Future Developments: It's important to acknowledge the challenges associated with these technologies, including cost, infrastructure requirements, and scalability. However, ongoing research and development are continually improving these energy-saving features and making them more accessible.

Consumer Adoption: Encouraging consumers to embrace energy-saving features, such as regenerative braking in electric vehicles, requires education and incentives. Governments and organizations are working to promote these technologies to create a more sustainable future. Remember to specify which aspect you'd like to explore further, and I can provide more detailed information on the challenges.

# Chapter 3:

## Overcoming Challenges

Overcoming the challenges of hybrid work environments can be a complex but essential task for organizations in today's evolving workplace landscape. The text provides several crucial points to consider.

Flexible Technology Infrastructure: Invest in technology that supports seamless communication and collaboration between in-office and remote workers. This may include cloud-based tools, video conferencing platforms, and secure data access.

Clear Communication:  Establish clear communication channels and guidelines for both remote and in-office employees. Regular team meetings, updates, and project management tools can help bridge the gap.

Cybersecurity: Protecting sensitive data becomes more critical in a hybrid setup. Implement robust cybersecurity measures to safeguard company information and ensure remote workers follow security protocols.

Employee Wellbeing: Recognize and address potential challenges to employee well-being, such as isolation for remote workers and burnout from excessive digital collaboration. Offer resources for mental health and encourage a healthy work-life balance.

Training and upskilling: Invest in training programs to help employees adapt to hybrid work. This might include digital skills, time management, and self-motivation strategies.

Performance Metrics: Develop fair and transparent performance metrics that account for the unique challenges of hybrid work. The focus should be on achieving desired outcomes rather than solely on the number of hours worked.

Company Culture: Nurture a strong company culture that connects both remote and in-office employees. Regular team-building activities, social events, and shared values can help foster a sense of belonging.

Feedback Loops: Create feedback mechanisms for employees to voice their concerns and ideas regarding the hybrid work model. Act on this feedback to continually improve the work environment.

Legal and Compliance: Ensure compliance with labor laws and regulations that may vary in different locations. Seek legal counsel to navigate the potential legal challenges associated with hybrid work.

Flexibility: Be adaptable and open to change. The hybrid work model may need adjustments as circumstances evolve, so stay responsive to employee needs and changing business dynamics. By addressing these challenges strategically, organizations can create a successful hybrid work environment that benefits both the company and its employees.

## Infrastructure and Adoption of Electronic and Hybrid Vehicles:

Infrastructure and Adoption of Electronic and Hybrid Vehicles The adoption of electronic and hybrid vehicles has been steadily increasing in recent years, driven by a combination of environmental concerns, government incentives, and advances in technology. This shift towards cleaner and more sustainable transportation options has significant implications for infrastructure development and the automotive industry as a whole. A key challenge in the adoption of electric vehicles (EVs) is the availability of

charging infrastructure: To encourage EV adoption, governments and private companies are investing in building a network of charging stations. There are three primary types of stations:

Home Charging: Most EV owners charge their vehicles at home using standard electrical outlets or dedicated home charging stations. This is the most convenient and cost-effective method for daily charging.

PublicCharging: Public charging stations are crucial for long-distance travel and urban charging needs. They can be found at various locations, including shopping centers, parking garages, and along highways.

Fast Charging: Fast-charging stations, often located along highways, provide a quick recharge for EVs, typically in under an hour. They are vital for reducing range anxiety and making long journeys feasible.

## Government Incentives: Many governments worldwide have introduced incentives to promote the adoption of electric and hybrid vehicles. The incentives offered include tax credits, rebates, and access to carpool lanes. Such policies help reduce the upfront cost of these vehicles and make them more attractive to consumers.

## Technological Advancements: Advances in battery technology have significantly improved the range and efficiency of electric vehicles. Lithium-ion batteries, for example, have become lighter, more powerful, and less expensive. These advancements are making electric vehicles a more practical choice for a broader range of consumers.

**Environmental Benefits:** Electric and hybrid vehicles produce fewer greenhouse gas emissions compared to traditional internal combustion engine (ICE) vehicles. As awareness of climate change and air quality issues grows, consumers are increasingly motivated to choose environmentally friendly transportation options.

**Automaker Commitment:** Major automakers are making substantial commitments to electrify their vehicle lineups. This shift towards electric and hybrid models is not only driven by consumer demand but also by stricter emissions regulations in many countries.

**Challenges:** Despite the upward momentum, difficulties still need to be surmounted.Range anxiety, the limited availability of certain models, and the need for more affordable EV options are some of the obstacles.The issue requires further attention for its wider adoption.

Hybrid Vehicles: Hybrid vehicles, which combine internal combustion engines with electric propulsion, offer a transitional solution for consumers who may not be ready to fully switch to electric vehicles. They provide improved fuel efficiency and lower emissions compared to traditional vehicles. In conclusion, the infrastructure and adoption of electronic and hybrid vehicles are closely intertwined. As charging infrastructure continues to expand, government policies incentivize adoption, and automakers invest in electric vehicle technology, we can expect to see an accelerated transition to cleaner and more sustainable transportation options in the coming years. This shift not only benefits the environment but also presents new opportunities and challenges for the automotive industry and society as a whole.

# Addressing concerns about limited charging infrastructure:

Addressing Concerns About Limited Charging Infrastructure As electric vehicles (EVs) gain popularity, concerns about limited charging infrastructure have become more prevalent. While the growth of EVs is undoubtedly a positive step toward reducing carbon emissions and dependence on fossil fuels, addressing charging infrastructure challenges is essential for widespread adoption. Here, we explore the concerns surrounding limited charging infrastructure and potential solutions:

Range Anxiety: One of the primary concerns for EV owners is the fear of running out of charge before reaching their destination. To alleviate this, expanding the charging network and deploying high-speed chargers along major highways are crucial. Additionally, better battery technology that offers longer ranges can mitigate this concern.

Urban Charging: In densely populated urban areas, access to home charging can be challenging due to limited parking spaces. To address this issue, cities can promote the installation of public charging stations in parking lots, curbsides, and residential areas. Incentives for apartment complexes to offer charging infrastructure can also help.

Rural Areas: Rural regions often lack charging stations, discouraging potential EV buyers. Expanding the charging network into these areas can be a challenge due to the lower population density. However, government incentives and public-private partnerships can make it more economically viable to establish charging stations in rural locations.

Charging Speed: Fast-charging stations are essential to reducing charging times. The development of more powerful chargers, like Tesla's Superchargers, and their widespread deployment can help alleviate concerns about lengthy charging stops during long trips.

Charging Standards: The existence of multiple charging standards (e.g., CCS, CHAdeMO, and Tesla's proprietary system) can confuse users and limit interoperability. Widespread adoption of standardized connectors and better cooperation among automakers can resolve this issue.

Grid Capacity: A sudden surge in EV adoption could strain the electrical grid. Upgrading the grid infrastructure and implementing smart charging solutions that distribute demand more evenly can mitigate these concerns.

Incentives: Governments can play a vital role in promoting EV adoption by offering incentives for installing charging infrastructure, both at home and in public places. Tax credits, subsidies, and grants can encourage businesses and individuals to invest in charging stations.

Education: Educating the public about charging options, the benefits of EVs, and how to locate charging stations can reduce anxiety and increase EV adoption rates.

Innovative Solutions: Companies are exploring innovative solutions like mobile charging units, which can be deployed temporarily to areas with limited infrastructure. Battery swapping stations are also being considered to provide quick battery replacements.

Collaboration: Collaboration among automakers, energy companies, and governments is essential. These stakeholders must work together to create a cohesive strategy for expanding and maintaining the charging system.

infrastructure:  In conclusion, while concerns about limited charging infrastructure for electric vehicles are valid, concerted efforts at local, national, and global levels can overcome these challenges. A comprehensive approach that includes infrastructure expansion, technological advancements, policy incentives, and public awareness campaigns will pave the way for a future where EVs are a practical and sustainable transportation option for all.

## Analyzing government policies and incentives to promote adoption

Analyzing government policies and incentives to promote adoption is a broad and multifaceted topic that can vary significantly by country and the specific context of adoption (e.g., technology adoption, renewable energy adoption, or adoption of certain behaviors).

Definition of Adoption: Define what you mean by "adoption" in your context. Is it the adoption of new technologies, renewable energy sources, healthcare practices, or something else?

Government Policies: Government policies play a crucial role in promoting adoption. These policies can include tax incentives, subsidies, grants, or regulatory frameworks that make it easier or more attractive for individuals or businesses to adopt the desired technology or behavior.

Government Incentives: Analyze the economic incentives provided by the government. For example, tax credits for businesses that invest in research and development or individuals who purchase energy-efficient appliances

Environmental Incentives: In cases related to sustainability and environmental conservation, governments often offer incentives to encourage the adoption of clean technologies, such as solar panels or electric vehicles, to reduce carbon emissions.

Education and Awareness: Government efforts to promote adoption may also include education and awareness campaigns to inform the public about the benefits of adopting certain technologies or practices.

Barriers to Adoption: Identify and analyze any barriers that might hinder adoption. These could include high upfront costs, a lack of infrastructure, or resistance to change.

Case Studies: Explore specific case studies or examples of government policies and their impact on adoption rates. This can provide concrete evidence of the effectiveness of certain incentives.

Public-Private Partnerships: Governments often collaborate with private-sector entities to promote adoption. Analyze how these partnerships work and their impact. Long-Term Sustainability: Consider whether the policies and incentives are sustainable in the long run or if they need to be adjusted as technologies evolve.

Measuring Success: Assess how success is measured in terms of adoption. Is it increased market share, reduced emissions, improved

public health, or some other metric?
Challenges and Criticisms: Be sure to address any challenges or criticisms associated with government policies and incentives. This could include concerns about fairness, unintended consequences, or the effectiveness of the incentives.

Global Perspectives: If relevant, compare and contrast government policies and incentives for adoption in different countries to provide a broader perspective. Remember to use credible sources, data, and research to support your analysis, as government policies and their impact can be complex and multifaceted.

## Discussing strategies to overcome range anxiety and improve the user experience:

Range anxiety is a common concern among electric vehicle (EV) owners, but there are several effective strategies to overcome it and enhance the overall user experience:

Improved Battery Technology: Advancements in battery technology continue to extend the range of EVs. Manufacturers are developing higher-capacity batteries that can store more energy and provide longer driving distances on a single charge.

Charging Infrastructure Expansion: Expanding the charging infrastructure is crucial. Governments and private companies are investing in building more charging stations, reducing the fear of running out of power while on the road.

Range Estimation Accuracy:
Manufacturers should improve the accuracy of range estimations displayed on EV dashboards. This can help drivers plan their trips more effectively and reduce anxiety.

Education and Information: Providing drivers with better information about EVs, including how factors like weather, driving habits, and terrain affect range, can empower them to make informed decisions and alleviate anxiety.

Mobile Apps and Navigation:
EV-specific navigation apps can help users find charging stations along their route, including their real-time availability and compatibility with their vehicle.

Battery Management Systems: Advanced battery management systems can optimize battery performance, extending its lifespan and maintaining a consistent range over time.

Flexible Charging Solutions: Home charging options, including Level 2 chargers, can offer convenience and peace of mind. Battery swapping services are also being explored in some regions.

Electric Range Extenders: Plug-in hybrid electric vehicles (PHEVs) combine a gasoline engine with an electric motor, providing a backup power source for longer trips and reducing range anxiety.

Community and peer support: Building a sense of community among EV owners can be valuable. Sharing experiences, tips, and advice through online forums and local EV clubs can help alleviate anxiety.

Resale Value Consideration: Manufacturers should ensure that EVs retain their resale value, so owners feel confident in their long-term investment.

Environmental Benefits: Emphasizing the environmental advantages of EVs can encourage users to focus on the positive aspects of their vehicles, helping reduce anxiety. In conclusion, overcoming range anxiety and enhancing the user experience with electric vehicles involves a combination of technological advancements, infrastructure development, and improved information and support systems. As these strategies continue to evolve, the transition to electric mobility will become more seamless and enjoyable for users.

# Chapter 4:

# Beyond Four Wheels:

Beyond Four Wheels" is a phrase often used to describe the world of mobility and transportation that goes beyond traditional automobiles. It signifies the evolution of transportation to include innovative and sustainable modes of travel. This can encompass various aspects:

Electric Vehicles (EVs): The transition from traditional internal combustion engine vehicles to electric vehicles is a significant part of this shift. EVs are seen as a more environmentally friendly option with the potential to reduce carbon emissions.

Autonomous Vehicles: The development of self-driving cars is another aspect. These vehicles have the potential to revolutionize transportation by increasing safety and efficiency.

Shared Mobility: Beyond owning a car, shared mobility services like ride-sharing, bike-sharing, and scooter-sharing are gaining popularity. This shift towards sharing resources can reduce the number of vehicles on the road and decrease congestion.

Urban Planning: The concept also extends to urban planning, where cities are redesigning their infrastructure to be more pedestrian and cyclist-friendly, with an emphasis on public transportation.

Sustainability:  Sustainability is a key driver of this movement. Beyond Four Wheels also encompasses the use of renewable energy sources to power transportation and reduce the environmental impact of travel.

Alternative Transportation: This includes modes such as electric scooters, electric skateboards, and e-bikes, which provide alternative and eco-friendly ways to get around cities.

Flying Vehicles: Some envision the future of transportation involving flying vehicles, like air taxis and drones, which could provide rapid point-to-point transportation in congested urban areas.

In summary, "Beyond Four Wheels" reflects the ongoing transformation of transportation into a more sustainable, efficient, and diverse ecosystem that includes electric vehicles, autonomous technology, shared mobility, and a focus on reducing environmental impact.

## The Future of Electronic and Hybrid Transportation:

The future of electronic and hybrid transportation is an exciting and evolving topic. It encompasses advancements in electric vehicles (EVs), hybrid technology, and sustainable transportation solutions.

Electric Vehicles (EVs): EVs are becoming increasingly popular due to their environmental benefits and advancements in battery technology. The future holds promises of longer ranges, faster charging times, and more affordable EV options.

Hybrid Technology: Hybrid vehicles, which combine traditional internal combustion engines with electric propulsion, continue to improve in terms of efficiency and performance. Plug-in hybrids are gaining popularity, offering the flexibility of both electric and gasoline power.

Autonomous Vehicles: Self-driving electric and hybrid vehicles are a significant part of the future. They have the potential to reduce accidents, traffic congestion, and energy consumption by optimizing routes and driving behavior.

Charging Infrastructure: The expansion of charging infrastructure is crucial for the widespread adoption of EVs. Fast-charging networks and innovative charging solutions will become more common.

Sustainable Transport: Beyond cars, the future of transportation includes electric buses, trucks, and even electric aircraft. Sustainable transportation is a key focus for reducing greenhouse gas emissions.

Environmental Impact: The shift toward electronic and hybrid transportation plays a vital role in reducing air pollution and greenhouse gas emissions, contributing to a cleaner and more sustainable future.

Government Initiatives: Governments worldwide are implementing policies to promote electronic and hybrid transportation. Incentives such as tax credits and rebates are encouraging consumers to make the switch.

Advances in battery technology:Are making electric vehicles (EVs) more accessible and efficient, with solid-state batteries potentially revolutionizing the industry.

Consumer Adoption: Conscious, the demand for electronic and hybrid vehicles increases, driving manufacturers to innovate.The increasing demand for electronic and hybrid vehicles among environmentally conscious consumers is driving manufacturers to innovate and offer more options.

Integration with Renewable Energy: Integrating EV charging with renewable energy sources like solar power is an exciting prospect for reducing the carbon footprint of transportation. The future of electronic and hybrid transportation is marked by sustainability, technological innovation, and a shift toward cleaner and more efficient modes of travel. It will be fascinating to witness how these developments continue to shape the way we move from place to place.

# Exploring developments in electronic and hybrid vehicles for public transportation:

Certainly! Exploring developments in electronic and hybrid vehicles for public transportation reveals several exciting trends and advancements:

## Electrification of Public Transit: Many cities worldwide are electrifying their public transit systems to reduce greenhouse gas emissions and improve air quality. Electric buses and trams are becoming increasingly common, offering quieter, cleaner, and more sustainable transportation options.

## Battery Technology Advancements: The development of high-capacity batteries and fast-charging infrastructure has been crucial for the success of electric public transportation. Lithium-ion batteries, solid-state batteries, and other emerging

technologies are enhancing the range and efficiency of electric vehicles.

Hybrid Powertrains: Hybrid buses and trains, which combine internal combustion engines with electric propulsion, are gaining popularity. These hybrids offer greater flexibility, especially for routes with longer distances or less charging infrastructure.

Autonomous Public Transport: Some cities are experimenting with autonomous electric buses and shuttles for public transportation. These autonomous vehicles can improve safety and efficiency while reducing labor costs.

Integration with Renewable Energy: Public transportation systems are increasingly powered by renewable energy sources, such as solar or wind power. This integration reduces the carbon footprint of electric and hybrid vehicles.

Improved Passenger Experience: Developments in interior design and amenities are enhancing the passenger experience on electric and hybrid public transport. Wi-Fi, USB charging ports, and comfortable seating are becoming standard features.

Smart Infrastructure: Public transportation systems are integrating smart technology to optimize routes, schedules, and passenger information. This helps reduce congestion and improve overall efficiency.

Environmental Benefits: The shift toward electronic and hybrid vehicles in public transportation aligns with the global push for sustainable mobility, reducing noise pollution and greenhouse gas emissions in urban areas.

Funding and Incentives: Governments and private companies are offering incentives and subsidies to encourage the adoption of electric and hybrid public transportation, making it more financially viable for cities.

Challenges: Despite these advancements, challenges remain, such as the initial high cost of electric vehicles, limited charging infrastructure in some regions, and the need for ongoing maintenance and training. In conclusion, the developments in electronic and hybrid vehicles for public transportation are a promising step toward more sustainable, efficient, and environmentally friendly urban mobility. As technology continues to advance and infrastructure improves, the future of public transportation looks increasingly electric and hybrid-powered.

## discussing the potential of electronic and hybrid technologies;

Electronic and hybrid technologies have revolutionized various industries, offering a myriad of possibilities and innovations. These technologies combine elements of traditional electronic components with newer, more advanced concepts, creating a synergy that opens up new horizons in fields such as transportation, energy, and healthcare. In the realm of transportation, hybrid technologies have gained immense popularity, especially in the automotive sector. Hybrid vehicles, which combine internal combustion engines with electric propulsion systems, offer improved fuel efficiency and reduced emissions. This innovation not only addresses environmental concerns but also lowers the cost of fuel for consumers. Moreover, electronic and hybrid technologies are transforming the energy landscape.

Advances in battery technology, combined with the integration of renewable energy sources, have led to the development of smart grids and energy storage solutions. These innovations are essential for the widespread adoption of clean energy and the reduction of our reliance on fossil fuels. In healthcare, electronic and hybrid technologies play a pivotal role in enhancing patient care and diagnostics. Wearable devices equipped with sensors can monitor vital signs and transmit real-time data to healthcare providers, enabling remote monitoring and early intervention. Additionally, the fusion of electronic health records and artificial intelligence facilitates more accurate diagnoses and treatment recommendations. Furthermore, electronic and hybrid technologies are poised to impact various other sectors, such as agriculture, aerospace, and manufacturing. They enhance precision and efficiency, offering the potential to address global challenges like food security, sustainable resource management, and increased productivity. In conclusion, electronic and hybrid technologies hold tremendous potential to shape our future. Their ability to merge the best of both worlds, combining traditional and cutting-edge concepts, paves the way for innovative solutions that can address some of the most pressing challenges of our time. As research and development continue to advance in these areas, we can expect even more groundbreaking applications that will transform the way we live, work, and interact with our environment.

# Fina Summary

"Embracing the Future" is a call to action that inspires people and organizations to seize the chances presented by the future. The future prospects and difficulties are highlighted in the text.This forward-thinking approach emphasizes adaptability, innovation, and the willingness to harness emerging trends and technologies. Whether in the realms of business, personal development, or societal progress, embracing the future is about fostering a mindset that embraces change, explores new horizons, and seeks to shape a brighter tomorrow. This title serves as a reminder that by embracing the future, we can unlock boundless potential and drive progress in our ever-evolving world.